The Hidden World of Bacteria

Understanding Microbial Life

Bhupen Thapa

Table of Contents

CHAPTER 1
Introduction to Bacteria

What are Bacteria?

Bacteria are single-celled microorganisms that play a vital role in our world. They are incredibly diverse and can be found in virtually every environment on Earth, from the depths of the ocean to the soil in our backyard. Despite their small size, bacteria have a huge impact on our lives, both positively and negatively.

One of the key characteristics of bacteria is their prokaryotic cell structure, which means they lack a nucleus and other membrane-bound organelles found in eukaryotic cells. This simple structure allows bacteria to reproduce rapidly, with some species able to double their population size in as little as 20 minutes. This rapid growth is one reason why bacteria are so prevalent in our environment.

Bacteria can be classified into different groups based on their shape, metabolism, and other traits. The three main shapes of bacteria are cocci (spherical), bacilli (rod-shaped), and spirilla (spiral-shaped). Each shape has its own unique characteristics and functions. In terms of metabolism, bacteria can be categorized as aerobic (requiring oxygen for growth), anaerobic (not requiring oxygen), or facultative anaerobes (able to grow with or without oxygen).

Despite their small size, bacteria play a crucial role in various ecosystems, performing essential functions such as decomposition, nitrogen fixation, and nutrient cycling. They are also used in various industrial processes, such as the production of antibiotics, food fermentation, and bioremediation. However, some bacteria can also cause diseases in humans, animals, and plants, making them important targets for medical research and treatment.

In conclusion, bacteria are fascinating microorganisms that are essential for life on Earth. Understanding their biology, diversity, and ecological roles is crucial for advancing our knowledge of microbiology and developing new strategies for controlling harmful bacteria while harnessing the beneficial ones. By studying the hidden world of bacteria, we can gain a deeper appreciation for the complexity and importance of these tiny organisms.

History of Bacteriology

Bacteriology, the branch of microbiology that focuses on the study of bacteria, has a rich and fascinating history that spans centuries. The history of bacteriology can be traced back to the 17th century when Antonie van Leeuwenhoek first observed microorganisms through a simple microscope. His discovery paved the way for further exploration into the world of bacteria and laid the foundation for the field of bacteriology.

In the 19th century, the germ theory of disease was proposed by Louis Pasteur and Robert Koch, two prominent figures in the field of bacteriology. This theory revolutionized the way we understand infectious diseases and led to significant advancements in public health and medicine. Koch's postulates, a set of criteria for establishing the causative relationship between a microorganism and a disease, became a cornerstone of bacteriological research.

The discovery of antibiotics in the early 20th century further revolutionized the field of bacteriology. Alexander Fleming's discovery of penicillin, the first antibiotic to be widely used in medicine, marked a turning point in the treatment of bacterial infections. The development of antibiotics has saved countless lives and continues to play a crucial role in modern medicine.

As our understanding of bacteria has evolved, so too has the field of bacteriology. Today, bacteriologists study a wide range of topics, from the role of bacteria in the environment to the mechanisms of antibiotic resistance. Advances in technology, such as next-generation sequencing and bioinformatics, have opened up new avenues of research and discovery in the field.

The history of bacteriology is a testament to the power of scientific inquiry and discovery. By studying bacteria, we have gained valuable insights into the complexity of microbial life and its impact on human health and the environment. As we continue to unravel the mysteries of the microbial world, bacteriology will remain a vital and dynamic field of study with far-reaching implications for medicine, public health, and beyond.

Importance of Studying Bacteria

Bacteria are microscopic organisms that play a crucial role in the environment, human health, and various industries. Studying bacteria is essential for understanding their impact on human life and the ecosystem. In this subchapter, we will explore the importance of studying bacteria and how it can help us better understand microbial life. One of the main reasons why studying bacteria is important is their role in human health. Bacteria can both harm and benefit humans, with some causing diseases while others are essential for our digestion and immune system. By studying bacteria, scientists can develop new treatments for bacterial infections and improve our overall health.

Furthermore, bacteria play a vital role in the environment by cycling nutrients, decomposing organic matter, and maintaining soil fertility. Understanding how bacteria interact with their environment can help us develop sustainable agricultural practices and protect ecosystems from pollution and degradation.

In addition to human health and the environment, studying bacteria is crucial for various industries such as biotechnology, food production, and pharmaceuticals. Bacteria have been used to produce antibiotics, enzymes, and other valuable products that benefit society. By studying bacteria, researchers can discover new microbial strains with unique properties that can be harnessed for industrial applications.

Overall, studying bacteria is essential for gaining a deeper understanding of microbial life and its impact on our world. By exploring the hidden world of bacteria, we can unlock new insights into human health, environmental sustainability, and industrial innovation. This subchapter will delve into the fascinating world of bacteria and showcase the importance of studying these microscopic organisms.

CHAPTER 2
Structure and Function of Bacteria

Cell Structure of Bacteria

In the subchapter "Cell Structure of Bacteria," we delve into the fascinating world of bacterial cells and explore their unique structures that enable them to thrive in a variety of environments. Understanding the cell structure of bacteria is essential for gaining insights into their functions, interactions, and pathogenicity.

At the core of every bacterial cell is the cytoplasm, a gel-like substance that houses the cell's genetic material and essential cellular machinery. Unlike eukaryotic cells, bacterial cells lack a true nucleus, and their genetic material is contained within a single circular chromosome. This streamlined design allows bacteria to replicate and adapt quickly to changing conditions.

One of the defining features of bacterial cells is their cell wall, which provides structural support and protection. The composition of the cell wall varies among different types of bacteria, with some species having a thick layer of peptidoglycan, while others have a thinner layer or lack a cell wall altogether. The cell wall also plays a crucial role in determining the shape of the bacterium, whether it be spherical, rod-shaped, or spiral.

In addition to the cell wall, many bacteria possess an outer membrane that further protects the cell from external threats. This outer membrane is composed of lipopolysaccharides and proteins that help regulate the movement of molecules into and out of the cell. This barrier also contributes to the pathogenicity of certain bacteria by allowing them to evade the immune system and resist antibiotics.

Within the cytoplasm, bacterial cells contain ribosomes, the molecular machines responsible for protein synthesis. These ribosomes are smaller than those found in eukaryotic cells, reflecting the simpler nature of bacterial cells. Bacteria also have specialized structures such as flagella for movement, pili for attachment to surfaces, and capsules for protection. By studying the cell structure of bacteria, we gain a deeper understanding of their biology and potential for both harm and benefit to human health.

Metabolism of Bacteria

Bacteria are incredibly diverse microorganisms that play essential roles in various ecosystems, including the human body. One crucial aspect of bacterial life is their metabolism, which involves the processes by which they obtain and utilize energy to carry out their cellular functions. Understanding the metabolism of bacteria is crucial for studying their behavior, growth, and interactions with other organisms.

Bacteria can be classified based on their metabolic pathways, with the most common being aerobic and anaerobic metabolism. Aerobic bacteria require oxygen to survive and obtain energy through the process of aerobic respiration, which involves the breakdown of glucose into carbon dioxide and water. In contrast, anaerobic bacteria can survive in the absence of oxygen and obtain energy through fermentation or anaerobic respiration, depending on the specific metabolic pathways they possess.

Another important aspect of bacterial metabolism is the ability of some bacteria to fix nitrogen, a process that involves converting atmospheric nitrogen into a form that can be used by plants and other organisms. This ability is essential for the nitrogen cycle, which is crucial for the growth of plants and the overall health of ecosystems. Bacteria that can fix nitrogen play a vital role in maintaining the balance of nutrients in the environment.

Bacteria also have diverse metabolic strategies for obtaining energy from various sources, including organic compounds, inorganic compounds, and sunlight. Some bacteria are heterotrophs, meaning they rely on organic molecules as a source of carbon and energy, while others are autotrophs, capable of synthesizing their organic molecules from inorganic sources. Photoautotrophic bacteria, such as cyanobacteria, can harness sunlight through photosynthesis to produce energy for their cellular processes.

Overall, the metabolism of bacteria is a complex and diverse process that allows these microorganisms to thrive in a wide range of environments. Understanding how bacteria obtain and utilize energy is crucial for studying their ecological roles, interactions with other organisms, and potential applications in biotechnology and medicine. By delving into the intricacies of bacterial metabolism, we can gain valuable insights into the hidden world of bacteria and their impact on the world around us.

Reproduction and Growth of Bacteria

Reproduction and growth are essential processes in the life cycle of bacteria. Bacteria are single-celled organisms that reproduce asexually through a process called binary fission. During binary fission, a single bacterial cell replicates its genetic material and divides into two identical daughter cells. This allows bacteria to rapidly multiply in a short period of time, leading to bacterial colonies that can grow exponentially.

The growth of bacteria is influenced by various factors, including environmental conditions such as temperature, pH, nutrients, and oxygen availability. Bacteria can be classified based on their optimal growth temperature into categories such as psychrophiles (cold-loving bacteria), mesophiles (moderate-temperature bacteria), and thermophiles (heat-loving bacteria). Different bacteria also have specific nutrient requirements for growth, with some being able to thrive in nutrient-rich environments while others can survive in nutrient-poor conditions.

Understanding the reproductive and growth patterns of bacteria is crucial for studying their behavior and controlling their populations. By knowing the conditions that favor bacterial growth, scientists can develop strategies to inhibit their growth and prevent the spread of harmful bacteria. This knowledge is especially important in the fields of medicine and public health, where controlling bacterial infections is a top priority.

In addition to binary fission, some bacteria also have the ability to exchange genetic material through a process called conjugation. During conjugation, two bacterial cells come into contact and transfer genetic material, allowing for the exchange of beneficial traits such as antibiotic resistance. This horizontal gene transfer mechanism plays a significant role in the evolution of bacteria and the development of antibiotic resistance.

Overall, the reproduction and growth of bacteria are complex processes that play a crucial role in the survival and adaptation of these microorganisms. By studying these processes, scientists can gain valuable insights into the behavior of bacteria and develop strategies to control their populations and prevent the spread of bacterial infections. This knowledge is essential for advancing our understanding of microbial life and improving public health outcomes.

CHAPTER 3
Bacterial Diversity

Classification of Bacteria

Bacteria are classified into different groups based on their characteristics, such as shape, size, and metabolism. This classification system helps scientists understand the diversity of bacterial species and their roles in various ecosystems. One common method of classifying bacteria is based on their shape, which can be spherical (cocci), rod-shaped (bacilli), or spiral (spirilla). This classification helps scientists identify and study different types of bacteria more easily.

Another important classification method for bacteria is based on their metabolism. Bacteria can be classified as either aerobic, which require oxygen to survive, or anaerobic, which do not require oxygen. Some bacteria are also classified based on their ability to produce energy through photosynthesis, while others rely on organic matter for energy. Understanding the metabolic capabilities of different bacteria is crucial for studying their roles in various environments.

Bacteria can also be classified based on their Gram staining characteristics. This method involves staining bacteria with a dye and observing their reaction to different chemicals. Bacteria that retain the dye and appear purple under a microscope are classified as Gram-positive, while bacteria that do not retain the dye and appear pink are classified as Gram-negative. This classification helps scientists differentiate between different types of bacteria and understand their cell wall structures.

In addition to shape, metabolism, and Gram staining characteristics, bacteria can also be classified based on their genetic makeup. Advances in DNA sequencing technology have allowed scientists to study the genetic composition of bacteria in more detail, leading to the identification of new bacterial species and their relationships to each other. This genetic classification system provides valuable insights into the evolutionary history of bacteria and their potential for genetic diversity.

Overall, the classification of bacteria is a complex and evolving field of study that continues to expand as new technologies and research methods are developed. By understanding the different classification systems used to categorize bacteria, scientists can gain a deeper insight into the diversity and complexity of microbial life. This knowledge is crucial for studying the roles of bacteria in various ecosystems, as well as for developing new strategies for combating bacterial infections and diseases.

Different Shapes and Sizes of Bacteria

Bacteria are incredibly diverse microorganisms that come in a wide range of shapes and sizes. Understanding the different shapes and sizes of bacteria is crucial for studying their behavior, functions, and interactions with their environment. In this subchapter, we will explore the various forms that bacteria can take, from simple spheres to complex spirals and everything in between.

One of the most common shapes of bacteria is the coccus, which are spherical or oval in shape. These bacteria can be found in clusters, chains, or pairs, and they can be either single-celled or form colonies. Examples of coccus-shaped bacteria include Streptococcus and Staphylococcus, which are known for causing a variety of infections in humans.

Another common shape of bacteria is the rod-shaped bacillus, which are elongated and cylindrical in shape. Bacillus bacteria can be found individually or in chains, and they are known for their ability to produce spores that allow them to survive in harsh environments. Examples of bacillus-shaped bacteria include Escherichia coli and Bacillus anthracis, which are both important in the fields of microbiology and medicine.

Spiral-shaped bacteria, or spirilla, are less common but still play important roles in various ecosystems. These bacteria have a helical shape and can be either rigid or flexible. Examples of spirilla-shaped bacteria include Helicobacter pylori, which is known for causing stomach ulcers, and Treponema pallidum, which causes syphilis. The unique shape of spirilla allows them to move efficiently through their environment and interact with other microorganisms.

Some bacteria have more complex shapes, such as filamentous bacteria that form long, branching chains. These bacteria are often found in soil and water environments, where they play important roles in nutrient cycling and decomposition. Examples of filamentous bacteria include Actinobacteria and Cyanobacteria, which are essential for maintaining healthy ecosystems and supporting plant growth.

In conclusion, the diversity of shapes and sizes of bacteria is a testament to their adaptability and versatility as microorganisms. By understanding the different forms that bacteria can take, researchers and healthcare professionals can better study and combat bacterial infections, as well as harness the beneficial properties of bacteria for various applications. As we delve deeper into the hidden world of bacteria, we continue to uncover the fascinating complexity of these microscopic organisms and their impact on the world around us.

Unique Characteristics of Bacteria

Bacteria are fascinating microorganisms with unique characteristics that set them apart from other forms of life. One of the most remarkable features of bacteria is their small size, typically ranging from 0.5 to 5 micrometers in length. Despite their tiny stature, bacteria are incredibly diverse in terms of shape, with some species appearing spherical, rod-shaped, spiral, or even filamentous.

Another distinctive trait of bacteria is their ability to thrive in a wide range of environments. From the depths of the ocean to the harsh conditions of hot springs, bacteria have adapted to survive in extreme habitats that would be uninhabitable for most other organisms. This adaptability is due in part to the diverse metabolic capabilities of bacteria, which allow them to utilize a variety of energy sources, including sunlight, organic matter, and even inorganic compounds.

Bacteria also possess a remarkable capacity for rapid reproduction, with some species able to divide every 20 minutes under ideal conditions. This high rate of reproduction contributes to the rapid evolution of bacteria, allowing them to quickly adapt to changes in their environment, including the presence of antibiotics or other stressors.

In addition to their small size, adaptability, and rapid reproduction, bacteria also exhibit a remarkable diversity of genetic material. Unlike more complex organisms, bacteria possess a single circular chromosome that contains all of their genetic information. In addition to their chromosome, bacteria may also contain plasmids - small, circular pieces of DNA that can be exchanged between individual bacteria, allowing for the rapid spread of beneficial traits.

Overall, the unique characteristics of bacteria make them a fascinating subject of study for scientists and researchers in the field of bacteriology. By understanding the diverse traits and capabilities of bacteria, we can gain valuable insights into the role that these microorganisms play in our world, from their contributions to nutrient cycling and ecosystem health to their potential applications in biotechnology and medicine.

CHAPTER 4
Bacterial Genetics

DNA Structure in Bacteria

The structure of DNA in bacteria plays a crucial role in their ability to survive and thrive in various environments. Unlike eukaryotic cells, which have a nucleus that houses their DNA, bacteria have a single circular chromosome located in the cytoplasm. This compact organization allows bacteria to replicate their DNA quickly and efficiently, enabling them to adapt to changing conditions and evolve rapidly.

The DNA in bacteria is made up of a double helix, consisting of two strands of nucleotides that are twisted around each other. These nucleotides contain the genetic information that dictates the traits and functions of the bacterium. The DNA molecule is stabilized by hydrogen bonds between complementary base pairs, which include adenine (A) pairing with thymine (T) and guanine (G) pairing with cytosine (C).

In addition to their main chromosome, bacteria may also contain plasmids, which are small circular pieces of DNA that can replicate independently of the main chromosome. Plasmids often carry genes that provide bacteria with specific advantages, such as antibiotic resistance or the ability to metabolize certain nutrients. This genetic flexibility allows bacteria to adapt to new challenges and exploit different resources in their environment.

The structure of DNA in bacteria is not static; it can be modified through processes such as mutation, recombination, and horizontal gene transfer. These mechanisms allow bacteria to acquire new genetic material from other bacteria or their environment, leading to genetic diversity and the emergence of new traits. This genetic plasticity is a key factor in the success of bacteria as a group, enabling them to survive in a wide range of habitats and outcompete other organisms.

Understanding the structure of DNA in bacteria is essential for studying their biology, evolution, and interactions with other organisms. By unraveling the genetic code of bacteria, scientists can gain insights into how these microorganisms function and how they can be harnessed for various applications, such as biotechnology and medicine. As we continue to uncover the hidden world of bacteria, the intricate structure of their DNA remains a central focus of research and discovery.

Genetic Variation in Bacteria

Genetic variation in bacteria is a fascinating topic that sheds light on the incredible diversity and adaptability of these microscopic organisms. Bacteria are known for their ability to rapidly evolve and develop resistance to antibiotics, making them a constant threat to human health. Understanding the mechanisms behind genetic variation in bacteria is crucial for developing effective strategies to combat bacterial infections. One of the primary ways in which bacteria acquire genetic variation is through horizontal gene transfer. This process allows bacteria to exchange genetic material with other bacteria, leading to the spread of beneficial traits such as antibiotic resistance. Horizontal gene transfer can occur through several mechanisms, including conjugation, transformation, and transduction. These processes play a key role in shaping the genetic diversity of bacterial populations.

Another important source of genetic variation in bacteria is mutation. Mutations can occur spontaneously during DNA replication, leading to changes in the genetic code of bacteria. While most mutations are harmful or neutral, some can provide bacteria with a survival advantage in certain environments. Over time, these advantageous mutations can become fixed in a bacterial population, contributing to genetic variation. The ability of bacteria to rapidly adapt to changing environments is due in large part to their high mutation rates and efficient mechanisms for genetic exchange. This genetic variation allows bacteria to evolve quickly in response to selective pressures, such as exposure to antibiotics or changes in environmental conditions. Understanding the genetic mechanisms behind bacterial adaptation is crucial for developing new treatments and strategies to combat bacterial infections.

In conclusion, genetic variation in bacteria is a complex and dynamic process that plays a crucial role in bacterial evolution and adaptation. By studying the mechanisms behind genetic variation in bacteria, researchers can gain valuable insights into how these organisms evolve and develop resistance to antibiotics. This knowledge is essential for developing effective strategies to combat bacterial infections and protect human health.

Horizontal Gene Transfer in Bacteria

Horizontal gene transfer in bacteria is a fascinating phenomenon that plays a crucial role in bacterial evolution and adaptation. Unlike vertical gene transfer, which occurs from parent to offspring, horizontal gene transfer involves the transfer of genetic material between different bacterial cells. This process allows bacteria to acquire new genes and traits that can enhance their survival in different environments.

There are three main mechanisms of horizontal gene transfer in bacteria: transformation, transduction, and conjugation. Transformation involves the uptake of naked DNA from the environment, while transduction involves the transfer of genetic material via bacteriophages, or viruses that infect bacteria. Conjugation, on the other hand, involves the direct transfer of genetic material between bacterial cells through a physical connection known as a pilus.

Horizontal gene transfer in bacteria can lead to the spread of antibiotic resistance genes, virulence factors, and other traits that can impact human health. For example, the transfer of antibiotic resistance genes between different bacterial species can make infections more difficult to treat with traditional antibiotics. Understanding the mechanisms of horizontal gene transfer is crucial for developing strategies to combat the spread of antibiotic resistance and other harmful traits in bacterial populations.

Horizontal gene transfer is not limited to bacteria within the same species, as genetic material can also be transferred between different species, genera, and even kingdoms. This process has significant implications for microbial diversity and evolution, as it allows bacteria to rapidly adapt to changing environmental conditions. By studying horizontal gene transfer in bacteria, scientists can gain insights into the mechanisms of evolution and the potential for genetic diversity in microbial populations.

In conclusion, horizontal gene transfer in bacteria is a complex and dynamic process that plays a key role in bacterial evolution and adaptation. By understanding the mechanisms of horizontal gene transfer, scientists can better comprehend how bacteria acquire new genes and traits, and how these processes can impact human health and the environment. Further research in this field is essential for developing strategies to combat the spread of antibiotic resistance and other harmful traits in bacterial populations.

CHAPTER 5
Bacterial Pathogenesis

Mechanisms of Bacterial Pathogenicity

Bacterial pathogenicity refers to the ability of certain bacteria to cause disease in humans, animals, and plants. Understanding the mechanisms behind this pathogenicity is crucial in developing effective strategies to prevent and treat bacterial infections. In this subchapter, we will delve into the various mechanisms that bacteria use to cause disease and evade the host immune system.

One key mechanism of bacterial pathogenicity is the production of virulence factors. These are molecules that enable bacteria to colonize and invade host tissues, evade immune responses, and cause damage to host cells. Examples of virulence factors include toxins, adhesins, and capsules. Toxins can directly damage host cells or disrupt normal cellular functions, while adhesins allow bacteria to attach to host tissues and colonize them. Capsules are protective structures that help bacteria evade the host immune system.

Another important mechanism of bacterial pathogenicity is the ability of bacteria to form biofilms. Biofilms are communities of bacteria that adhere to surfaces and produce a protective matrix of extracellular polymeric substances. Biofilms allow bacteria to resist antibiotics and immune responses, making infections difficult to treat. Biofilm formation is a common feature of many chronic bacterial infections, such as those caused by Pseudomonas aeruginosa and Staphylococcus aureus.

Bacteria can also manipulate the host immune system to their advantage. Some bacteria produce molecules that inhibit the host immune response, allowing them to evade detection and clearance by the immune system. Other bacteria can trigger an exaggerated immune response, leading to inflammation and tissue damage. Understanding how bacteria interact with the host immune system is essential for developing therapies that target specific immune pathways to combat bacterial infections.

In addition to these mechanisms, bacteria can also acquire antibiotic resistance genes through horizontal gene transfer. This allows them to survive in the presence of antibiotics and makes infections more difficult to treat. The spread of antibiotic-resistant bacteria is a major public health concern, highlighting the importance of developing new antibiotics and alternative treatment strategies. By understanding the mechanisms of bacterial pathogenicity, we can better combat bacterial infections and prevent the emergence of antibiotic resistance.

Host-Pathogen Interactions

In the fascinating realm of bacteriology, one of the key areas of study is host-pathogen interactions. This subchapter delves into the intricate dynamics between bacteria and their hosts, shedding light on the complex relationships that exist at the microscopic level. Understanding these interactions is crucial for developing effective strategies to combat infectious diseases and improve public health outcomes.

At the heart of host-pathogen interactions is the battle for survival between bacteria and their hosts. Pathogenic bacteria have evolved a myriad of strategies to evade the immune system and colonize host tissues, leading to infection and disease. Conversely, hosts have developed intricate defense mechanisms to recognize and eliminate invading pathogens, ranging from physical barriers to immune responses. This ongoing arms race between bacteria and hosts shapes the outcome of infections and influences disease severity.

One key concept in host-pathogen interactions is virulence, which refers to the ability of a pathogen to cause disease in a host. Virulence factors are molecular weapons that bacteria use to manipulate host cells and tissues, allowing them to establish infections and evade immune surveillance. Understanding the mechanisms by which bacteria become virulent is crucial for developing targeted therapies to combat infectious diseases and reduce the burden of antibiotic resistance.

In addition to virulence factors, host-pathogen interactions are also influenced by the host's immune system. The immune response plays a critical role in defending the host against invading pathogens, through a complex interplay of cells and molecules that work together to identify and eliminate foreign invaders. Dysregulation of the immune response can lead to either inadequate protection, resulting in chronic infections, or excessive inflammation, leading to tissue damage and sepsis.

By unraveling the intricacies of host-pathogen interactions, bacteriologists can gain valuable insights into the mechanisms of infectious diseases and develop novel approaches for prevention and treatment. This subchapter serves as a window into the hidden world of bacteria, highlighting the ongoing battle between pathogens and hosts that shapes the landscape of infectious diseases. Through continued research and collaboration, we can hope to unravel the mysteries of host-pathogen interactions and pave the way for a healthier future.

Strategies for Controlling Bacterial Infections

Bacterial infections can range from mild nuisances to life-threatening conditions if left unchecked. In order to prevent and control these infections, it is essential to understand the strategies that can be employed to combat bacterial growth and spread. This subchapter will explore some of the most effective methods for controlling bacterial infections.

One of the most fundamental strategies for controlling bacterial infections is proper hygiene. By washing hands regularly with soap and water, individuals can reduce the risk of spreading bacteria from one surface to another. Additionally, practicing good hygiene habits such as covering coughs and sneezes can help prevent the transmission of harmful bacteria.

Another important strategy for controlling bacterial infections is the use of antibiotics. Antibiotics are medications that work by killing or inhibiting the growth of bacteria. It is important to use antibiotics only as prescribed by a healthcare professional, as misuse or overuse can lead to antibiotic resistance, making it more difficult to treat bacterial infections in the future.

In addition to antibiotics, vaccines are an important tool for controlling bacterial infections. Vaccines work by stimulating the body's immune system to produce antibodies against specific bacteria, providing immunity against future infections. By ensuring that individuals receive recommended vaccinations, the spread of bacterial infections can be greatly reduced.

Finally, environmental control measures can also play a key role in preventing bacterial infections. This can include measures such as disinfecting surfaces, proper waste disposal, and ensuring clean water sources. By implementing these strategies, individuals can help to create a safer and healthier environment that is less hospitable to harmful bacteria.

CHAPTER 6
Applications of Bacteriology

Industrial Uses of Bacteria

Bacteria play a crucial role in various industrial processes due to their ability to carry out specific biochemical reactions. One of the most well-known industrial uses of bacteria is in the production of antibiotics. Bacteria such as Streptomyces are used to produce antibiotics like penicillin and tetracycline, which are essential for treating bacterial infections in humans. These antibiotics are produced through fermentation processes, where bacteria are grown in large-scale bioreactors and their byproducts are harvested for pharmaceutical use.

Another important industrial use of bacteria is in the production of enzymes. Enzymes are proteins that catalyze chemical reactions, and bacteria produce a wide range of enzymes that are used in various industries. For example, bacteria like Bacillus are used to produce enzymes that are used in laundry detergents to break down tough stains. Other bacteria are used to produce enzymes that are used in food processing, waste treatment, and biofuel production.

Bacteria are also used in the production of bioplastics. Bioplastics are a sustainable alternative to traditional plastics that are made from petroleum. Bacteria like Pseudomonas are used to produce bioplastics through fermentation processes. These bioplastics are biodegradable and can be used in a wide range of applications, from packaging to medical devices.

In addition to antibiotics, enzymes, and bioplastics, bacteria are also used in the production of vitamins and amino acids. Bacteria like Corynebacteria and Escherichia coli are used to produce vitamins like vitamin B12 and amino acids like lysine. These vitamins and amino acids are used in various industries, including food and pharmaceuticals, to fortify products and enhance nutritional value.

Overall, bacteria play a crucial role in various industrial processes, from the production of antibiotics to enzymes, bioplastics, vitamins, and amino acids. By harnessing the unique biochemical capabilities of bacteria, industries can produce a wide range of products that are essential for human health, sustainability, and innovation. Understanding the industrial uses of bacteria is key to unlocking the full potential of these microscopic organisms in the modern world.

Environmental Impact of Bacteria

Bacteria play a crucial role in the environment, shaping ecosystems and influencing various processes. The environmental impact of bacteria is vast and multifaceted, with both positive and negative consequences. Understanding how bacteria interact with their surroundings is essential for managing ecosystems and addressing environmental challenges.

One of the most significant ways bacteria impact the environment is through their role in nutrient cycling. Bacteria are essential for breaking down organic matter, releasing nutrients such as nitrogen, phosphorus, and carbon back into the soil. This process is crucial for plant growth and overall ecosystem health. Without bacteria, nutrients would remain locked up in dead organic matter, limiting their availability to other organisms.

However, some bacteria can also have negative environmental impacts. For example, certain species of bacteria are responsible for causing diseases in plants, animals, and humans. Pathogenic bacteria can disrupt ecosystems by causing widespread illness and death among populations of organisms. Understanding the mechanisms by which pathogenic bacteria spread and cause harm is essential for preventing and managing disease outbreaks.

In addition to nutrient cycling and disease transmission, bacteria also play a role in environmental pollution. Some bacteria are capable of breaking down pollutants such as oil, pesticides, and heavy metals, helping to clean up contaminated sites. However, other bacteria can contribute to pollution by releasing harmful toxins or byproducts into the environment. Managing the presence of these bacteria is critical for reducing pollution and protecting environmental health.

Overall, the environmental impact of bacteria is complex and multifaceted. By studying how bacteria interact with their surroundings, researchers can gain valuable insights into ecosystem dynamics and environmental processes. Understanding the role of bacteria in nutrient cycling, disease transmission, and pollution can help us better manage and protect our environment for future generations.

Medical Advances in Bacteriology

Medical advances in bacteriology have revolutionized the field of medicine, leading to significant breakthroughs in the understanding and treatment of bacterial infections. One of the most important advancements in recent years is the development of rapid diagnostic tests for identifying bacterial pathogens. These tests allow healthcare providers to quickly identify the specific bacteria causing an infection, enabling them to prescribe the most effective treatment in a timely manner.

Another major advancement in bacteriology is the development of new antibiotics and antimicrobial agents. These drugs are designed to target specific bacterial pathogens, reducing the risk of antibiotic resistance and improving treatment outcomes for patients. Researchers continue to explore new ways to combat antibiotic resistance, including the development of novel antimicrobial compounds and alternative treatment strategies.

Along with advances in diagnostics and treatment options, researchers have made significant progress in understanding the mechanisms by which bacteria cause disease. By studying the interactions between bacteria and the human immune system, scientists have gained valuable insights into the pathogenesis of bacterial infections. This knowledge is essential for developing new vaccines and therapeutics to prevent and treat bacterial diseases.

In addition to advancements in treatment and prevention, researchers have also made strides in the field of bacteriology through the use of cutting-edge technologies such as genomics and proteomics. These techniques allow scientists to study the genetic and protein profiles of bacteria, providing valuable information about their virulence factors and antibiotic resistance mechanisms. By leveraging these technologies, researchers can better understand how bacteria evolve and adapt to their environment, leading to new strategies for controlling and combatting bacterial infections.

Overall, the medical advances in bacteriology have had a profound impact on healthcare, leading to improved diagnosis, treatment, and prevention of bacterial infections. As researchers continue to make progress in understanding microbial life, it is clear that the field of bacteriology will continue to play a critical role in shaping the future of medicine.

CHAPTER 7
Future Perspectives in Bacteriology

Emerging Trends in Bacterial Research

In recent years, bacterial research has seen a surge in interest and new discoveries, leading to the emergence of several exciting trends in the field. These trends are shaping the way we understand microbial life and are opening up new avenues for research and innovation. In this subchapter, we will explore some of the key emerging trends in bacterial research that are revolutionizing our understanding of these tiny but powerful organisms.

One of the most significant trends in bacterial research is the increasing focus on the human microbiome. The human microbiome is the collection of trillions of bacteria that inhabit our bodies, playing a crucial role in our overall health and wellbeing. Researchers are now exploring how these bacteria interact with our bodies and how they can be manipulated to improve our health. This research has led to groundbreaking discoveries in the fields of microbiology and immunology, and is paving the way for new treatments for a wide range of diseases.

Another important trend in bacterial research is the rise of antibiotic resistance. Antibiotic resistance is a growing global health crisis, with many bacteria becoming resistant to the drugs that were once used to treat them. Researchers are now focusing on developing new antibiotics and alternative treatments to combat this threat. This research is vital for ensuring that we have effective treatments for bacterial infections in the future and is driving innovation in the field of bacteriology.

Advances in technology have also revolutionized bacterial research, allowing scientists to study these organisms in ways that were previously impossible. Techniques such as next-generation sequencing and metagenomics have allowed researchers to study entire bacterial communities and uncover new insights into their biology and behavior. These technological advancements are helping to push the boundaries of our understanding of bacterial life and are driving new discoveries in the field.

In addition to these trends, researchers are also exploring the potential of using bacteria in biotechnology and environmental applications. Bacteria have incredible metabolic capabilities that can be harnessed for a wide range of purposes, from producing biofuels to cleaning up environmental contaminants. By studying the unique properties of bacteria and

19

manipulating their genetic material, researchers are unlocking the potential of these organisms for a variety of practical applications. This research is not only expanding our understanding of bacterial life, but also opening up new possibilities for using bacteria to address some of the biggest challenges facing society today.

Overall, the emerging trends in bacterial research are reshaping the field of bacteriology and leading to exciting new discoveries. By exploring the human microbiome, combating antibiotic resistance, leveraging technology, and harnessing bacteria for biotechnological and environmental applications, researchers are pushing the boundaries of our understanding of microbial life and paving the way for a future where bacteria play a central role in addressing some of the most pressing issues facing humanity.

Challenges and Opportunities in Bacteriology

Bacteriology is a field of study that focuses on the identification, classification, and study of bacteria. This subchapter will discuss the challenges and opportunities that researchers in bacteriology face in their quest to understand the hidden world of bacteria. Despite the challenges, there are also numerous opportunities for advancement and discovery in this exciting field.

One of the major challenges in bacteriology is the vast diversity of bacteria species. There are estimated to be trillions of different species of bacteria, many of which have not yet been discovered or studied. This presents a significant challenge for researchers who are trying to classify and understand the complex world of bacteria. Additionally, many bacteria are difficult to culture in a laboratory setting, making it challenging to study their characteristics and behavior.

Another challenge in bacteriology is the rise of antibiotic resistance. As bacteria evolve and develop resistance to antibiotics, it becomes increasingly difficult to treat infections. This has led to a growing public health crisis, as common bacterial infections become more difficult to treat. Researchers in bacteriology must work to develop new antibiotics and find alternative treatment methods to combat this growing threat.

Despite these challenges, there are also numerous opportunities in bacteriology. One of the biggest opportunities lies in the field of biotechnology. Bacteria are used in a variety of biotechnological applications, from producing antibiotics to cleaning up environmental pollution. Researchers in bacteriology have the opportunity to explore the potential of bacteria in a wide range of industries and applications.

Furthermore, advancements in technology have opened up new opportunities for researchers in bacteriology. Techniques such as next-generation sequencing and metagenomics allow researchers to study bacteria at a level of detail that was previously impossible. This has led to a greater understanding of the role that bacteria play in various ecosystems, as well as their potential applications in medicine, agriculture, and industry.

In conclusion, bacteriology is a field that presents both challenges and opportunities for researchers. By overcoming the challenges of bacterial diversity and antibiotic resistance, researchers have the opportunity to make significant advancements in our understanding of bacteria and their potential applications. With the continued development of technology and research methods, the field of bacteriology is poised for exciting discoveries and breakthroughs in the years to come.

The Role of Bacteriology in Shaping the Future of Medicine

Bacteriology plays a crucial role in shaping the future of medicine by providing valuable insights into the world of microbes. As our understanding of bacteria continues to evolve, we are discovering new ways to combat infectious diseases, develop antibiotics, and improve public health. By studying the structure, physiology, and behavior of bacteria, scientists can better understand how these tiny organisms impact human health and the environment.

One of the key contributions of bacteriology to medicine is the development of antibiotics. These powerful drugs are essential for treating bacterial infections and have saved countless lives since their discovery. By studying the mechanisms of bacterial resistance to antibiotics, researchers can develop new drugs that are more effective against resistant strains. This ongoing research is crucial for staying one step ahead of evolving bacteria and preventing the spread of drug-resistant infections.

In addition to antibiotic development, bacteriology also plays a vital role in understanding the role of bacteria in human health. The human microbiome, which consists of trillions of bacteria living in and on our bodies, plays a crucial role in maintaining our overall health. By studying the interactions between bacteria and the human immune system, scientists can develop new treatments for conditions such as inflammatory bowel disease, obesity, and autoimmune disorders.

Furthermore, bacteriology is essential for monitoring and controlling infectious diseases. By studying the epidemiology of bacterial infections, scientists can track the spread of diseases and develop strategies for prevention and control. This knowledge is crucial for responding to outbreaks of infectious diseases and protecting public health. Through the study of bacteriology, we can better understand the factors that contribute to the emergence and spread of infectious diseases and develop effective interventions to prevent future outbreaks.

In conclusion, bacteriology is a foundational science that plays a critical role in shaping the future of medicine. By studying the world of bacteria, researchers can develop new antibiotics, understand the human microbiome, and control infectious diseases. As our knowledge of bacteriology continues to grow, we can look forward to a future where we are better equipped to combat bacterial infections, improve human health, and protect public health.

CHAPTER 8
Conclusion

Summary of Key Points

In this subchapter, we have explored the fascinating world of bacteria and the crucial role they play in our lives. From their microscopic size to their incredible diversity, bacteria are truly remarkable organisms. We have learned that bacteria can be found everywhere, from the soil beneath our feet to the depths of the ocean. They are essential for many processes, such as nutrient cycling and decomposition, and they also play a key role in human health, both as pathogens and as beneficial symbionts.

One key point to remember is that not all bacteria are harmful. In fact, many bacteria are beneficial to humans and other organisms. For example, certain bacteria in our gut play a crucial role in digestion and help protect us from harmful pathogens. Understanding the diversity of bacteria and their different roles is essential for harnessing their potential for human health and environmental sustainability.

Another important point to consider is the impact of human activities on bacterial populations. Pollution, antibiotic overuse, and habitat destruction can all have negative effects on bacterial communities, leading to imbalances that can have far-reaching consequences. By studying bacteria and their interactions with the environment, we can better understand how to protect and promote the health of both microbial communities and the ecosystems they inhabit.

It is also important to recognize the potential dangers posed by pathogenic bacteria. From foodborne illnesses to antibiotic-resistant infections, pathogenic bacteria can pose serious threats to human health. Understanding how these bacteria spread and cause disease is essential for developing effective strategies to prevent and treat infections. By studying the mechanisms of bacterial pathogenesis, researchers can develop new ways to combat these dangerous pathogens and protect public health.

In conclusion, bacteria are incredibly diverse and important organisms that play a crucial role in the health of both humans and ecosystems. By understanding the key points discussed in this subchapter, we can gain a deeper appreciation for the hidden world of bacteria and the vital role they play in our lives. Whether studying the beneficial bacteria in our guts or working to combat dangerous pathogens, the study of bacteria is essential for advancing our understanding of microbial life and improving human health.

Final Thoughts on the Hidden World of Bacteria

In conclusion, the hidden world of bacteria is a fascinating and complex realm that plays a crucial role in our daily lives. From aiding in digestion to helping us fight off infections, bacteria are essential for our survival. By understanding more about these microorganisms, we can better appreciate their importance and work towards harnessing their potential for the benefit of humanity.

One of the key takeaways from exploring the hidden world of bacteria is the importance of maintaining a healthy balance of bacteria in our bodies. Disruptions to this delicate balance can lead to a range of health issues, from digestive problems to infections. By taking steps to support our microbiome through a healthy diet, probiotics, and other interventions, we can help ensure that our bodies remain in harmony with these essential microorganisms.

Another important aspect of the hidden world of bacteria is the role that they play in the environment. Bacteria are essential for breaking down organic matter, recycling nutrients, and maintaining the balance of ecosystems. By understanding more about how bacteria interact with their environment, we can work towards better stewardship of our planet and ensure that these vital microorganisms continue to thrive.

As our understanding of bacteria continues to evolve, it is becoming increasingly clear that these tiny organisms hold vast potential for medical and scientific advancements. From new antibiotics to innovative biotechnologies, bacteria are proving to be invaluable resources for addressing some of the most pressing challenges facing humanity today. By continuing to explore and unlock the secrets of the hidden world of bacteria, we can pave the way for a healthier, more sustainable future.

In closing, the hidden world of bacteria is a rich and diverse realm that holds countless secrets waiting to be discovered. By delving deeper into the mysteries of these microorganisms, we can gain a greater appreciation for their importance and unlock new opportunities for scientific and medical breakthroughs. As we continue to explore the hidden world of bacteria, let us approach this fascinating realm with curiosity, respect, and a sense of wonder at the incredible complexity of the microbial world.

Dear Reader,

Thank you for choosing "The Hidden World of Bacteria: Understanding Microbial Life." I hope you found this exploration into the microscopic world of bacteria as fascinating and enlightening as I intended.

Your feedback is incredibly valuable to me and to other potential readers. If you enjoyed the book or found it helpful, please consider leaving a review on Amazon. Your review can help others discover the importance and wonder of bacterial life.

To leave a review, simply follow these steps:

1. Go to the book's page on Amazon.
2. Scroll down to the "Customer Reviews" section.
3. Click on "Write a customer review."
4. Share your thoughts and experiences with the book.

Thank you for your time and support. Your review means the world to me!

Best regards,

Bhupen Thapa